Speaker Recognition

Bandar Hezam

Bibliographic information published by the German National Library:

The German National Library lists this publication in the National Bibliography; detailed bibliographic data are available on the Internet at http://dnb.dnb.de.

ISBN: 9783346980236
This book is also available as an ebook.

Table of Contents

Table of Figures

1   Introduction.

The ability to recognize people by their voice is an important social behavior. Identifying a person based on speech alone, also, is known that speech is like a speaker-dependent feature that enables us to identify or recognize friends over the phone. Human voice comprises of numerous discriminative features with the ability to identify speakers, our voice comprises of significant energy of frequency range from zero to around 5kHz whereas the primary target of this assignment speaker identification is to extract, characterize and recognize individual speaker using an audio track of minimum of minute recording in order to identify information about the speaker identity Though temporal period of speech signals like correlation, zero crossing and lots more are assume constant over short period (Vibha Tiwari, 2010) that means, in the case of using hamming window, voice signal is divided into number of blocks of short duration in order to be able to do further processing such as normal Fourier series transform. However, speaker voice identification for audio track with group people can be challenging if individual voice activities are not properly analyzed or detected, to approach this issue voice activities detection (VAD) algorithm can approach, this is a very useful technique that helps to improve the performance of voice recognition or speaker identification system working in these scenarios. Voice activity detection (VAD) is used in most voice recognition systems within the features extraction process for speech enhancement. The techniques usually involve when noise statistics such as spectrum are estimated during the non-voice period so that the speech enhancement algorithm can be applied such as wiener filer or spectral subtraction. VAD is also useful for non-voice frame-dropping in voice recognition in order to reduce the number of insertion errors caused by noise. Therefore, this assignment aims to perform speaker identification from a group of people in a recorded audio track utilizes voice activities detection before performing the speaker identification. The voice activity detection plays an essential role in this assignment as it improves speech intelligibility and recognition, both speaker identification, and voice activity detection utilize the Mel Frequency Cepstrum Coefficient (MFCC) for voice feature extraction. Mel frequency Cepstrum Coefficient (MFCC) in this assignment is to extract speech features of the speaker and quantized to several centroids by utilizing a vector quantization algorithm (Vibha Tiwari, 2010). Therefore, it becomes important to research and develop a system that can allow us to identify the speaker through their voice.

1.1   Theoretical Concepts

Voice recognition is a computer software program or hardware device with the ability to decode the human voice. Voice recognition is a system that allows for a secure method of authenticating speakers, the system work in such a way that it general speaker model during the enrollment phase which based on the speaker characteristics. The system testing phase typically involves making a claim on the identity of an unknown speaker using the given speech characteristics and the trained models. However, speaker identification (Zilovic et al., 2014) is known to be one among the two categories of speaker recognition system because speaker recognition can be categorized also as speaker verification whereas, the main difference between both speaker identification and speaker verification ensure to known if the person speaking and claim to be is fully verified while speaker identification make multiple decision by comparing of the person speaking with the one trained or store in database as an attempt to identify the speaker. The interest of the assignment is speaker identification; therefore, speaker identification is the main focus for this study.

voice identification can be categorized into two main categories, these are text-independent and text-dependent voice identification (Fu Zhonghua and Zhao Rongchun, 2003). Both text-dependent voice identification and text-independent voice identification are different because one depends on a specific word and the others don't. therefore, the assignment will consider text-independent which means the system won't depend on specific words. Verification.

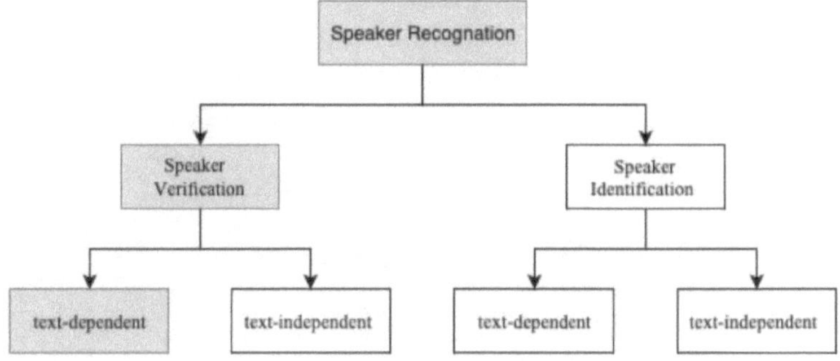

Figure 1:Speaker Recognition Theoretical Scope of Study.

1.1.1 Speaker Recognition

The theoretical analysis of speaker recognition is to ensure that, the system automatically recognizes the person that is a speaker or different speakers include the speech waves. This comprises two major sessions, the first session is known as a train or enrollment session while the second session refers to the testing or operation session. The system involves both sessions to provide a voice sample so that the trained model for the speaker.

1.1.2 Classification of Automatic Speaker Recognition

Automatic speaker recognition is often considered as one of the most economical and natural method for an authentication system, the major challenges in speaker recognition is the study and analysis of the speech signal characteristics and this makes the system unique among another type of signal. There are unique in every speech signal needed to analyze for identification and verification. In a situation when an individual recognizes a familiar voice and able to match the speaker's name to the voice recognition is known as speaker identification, however, the further process to validate who the user claim to be is known as verification. Speaker identification happens to us all the time also verification does occur to us as well.

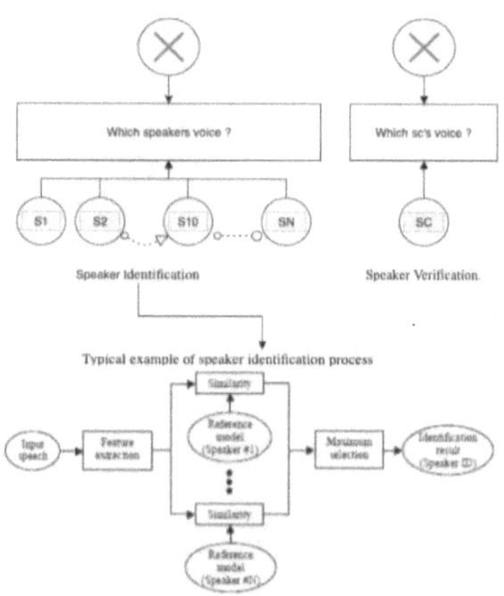

Figure 2:Speaker Identification and Speaker Verification.

1.1.3 Speech Feature Extraction

In speaker identification system, there are two modules involve, these are feature extraction and feature matching, feature extraction involves the procedure of extracting a small amount of data from voice signal that can use to represent each speaker while final part which is feature matching is the process of identifying the unknown speaker by comparing extracted features from the original voice input with the ones from a set of known speakers. There is a different method of voice feature extraction, some are stated as follows (Urmila Shrawankar, 2013.)

- Mel-frequency cepstral coefficients (MFCC)
- Linear predictive analysis (LPC)
- Linear predictive cepstral coefficients (LPCC)
- Mel scale cepstral analysis (MEL)
- Power spectral analysis (FFT)
- First order derivative (DELTA)

Linear predictive analysis (LPC) :

LPC is known to be one of the powerful speech analyses and very useful for encoding quality speech at a low bit rate. The main idea of linear predictive analysis is that a specific voice sample at the current time can be approximated as a linear combination of the past voice sample. The principle of LPC is majorly designed to minimize sue of the square difference between estimated speech signal and original speech signal over the finite duration. This provides a unique set of predictor coefficients and this estimated at every frame which usually takes 20ms long. The predictor coefficient can be represented as ak. In this feature extraction techniques, another major parameter is gain (G).

$$H(z) = G/(1-\Sigma akz\text{-}k)$$

Where k=1 to p, this is 10 for the LPC-10 algorithm and 18 for the improved algorithm that is utilized.

Furthermore, the auto-correlation method is to computer with Levinsion-Durbin recursion (Deller et al., 2000). However, LPC performance analysis can be stated as follow

- Computational Complexity
- Overall Delay of the System
- Bit Rates
- Objective Performance Evaluation

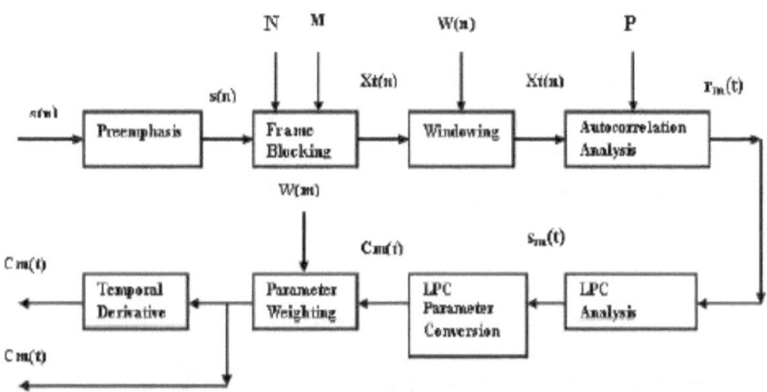

Figure 3:The block diagram describes LPC processor.

Mel-frequency cepstral coefficients (MFCC)

The utilization of Mel Frequency Cepstral Coefficients is known to be one of the most standard methods for voice feature extraction (Motlíček, 2002). The utilization of the 20 MFCC coefficient is very common in ASR, however, the 10-12 coefficient most considered to be sufficient enough for coding speech (Hagen at al., 2015). However, MFCC downside is the noise sensitivity due to its dependence on the spectral form and to overcome this problem, periodicity of speech signals could be used (Ishizuka and Nakatani, 2006).

Editor's note:
This image was removed for
copyright reasons.

Figure 4:Pipeline MFCC.

The Non-linear frequency scale utilizes an approximation to the Mel-frequency scale, this approximate linear frequency below 1 kHz along with a logarithm for frequency above 1 kHz. The idea is motivated because the human auditory system becomes less frequency-selective as a frequency increase above 1 kHz. Therefore, MFCC features typically correspond to the cepstrum of the filter-bank energies and can generally calculate first by computing the log energy from the filter-bank output with the following equation.

$$S_t[m] = In\left(\sum_{n=0}^{N-1} |x_t[n]|^2\, H_m[n]\right) \quad 0 \leq m < M$$

Where:

Xt[n] is known as DFT of the tth input speech frame,

Hm[n] is known as the frequency response of mth filter in the filter-bank,

N is known as the window size of the transform

M become the total number of filters.

After the above equation, we can now computer the discrete cosine transform (DCT) of the log energies with the following equation.

$$\overrightarrow{C_t}[m] \sum_{n=0}^{M-1} S_t[m] \cos \left(\frac{n - 0.5}{M} \right) \quad 0 \leq m < M$$

However, using the techniques that human auditory system is sensitive to tome evolution of the spectral signal made it possible for the extraction of this information as part of feature analysis, therefore two different first and second-order can be utilized to capture the changes in the coefficient over time and computed respectively as follow.

$$\Delta \overrightarrow{C_t} = \Delta \overrightarrow{C_t} + 2 - \overrightarrow{C_t} - 2$$
$$\Delta \overrightarrow{C_t} = \Delta \overrightarrow{C_t} + 1 - \overrightarrow{C_t} - 1$$

Therefore, these dynamic coefficients can be concatenated with the static coefficients, to make up the final output of feature analysis representing the tth speech frame, the following equation can be utilized.

$$\overrightarrow{x_t} = \begin{bmatrix} \overrightarrow{C_t} & \Delta\overrightarrow{C_t} & \Delta\Delta\overrightarrow{C_t} \end{bmatrix}^T$$

2   Objectives

In order to meet the requirements that have been set for this assignment, a research on the possible algorithms for speaker identification and be able to select one algorithm to be implemented with justification and to produce individual voice biometric of ten chosen speakers that will be used to develop appropriate solution for speaker identification using an unseen test record.

3 Design implementation.

In this section the algorithms utilize to implement and process the speaker identification process explain, the include major equation along with the programming code, this section begin with designed block diagram of the proposed speaker identification system which explain major procedure, with work also include the system flow chat and GUI design and explanation.

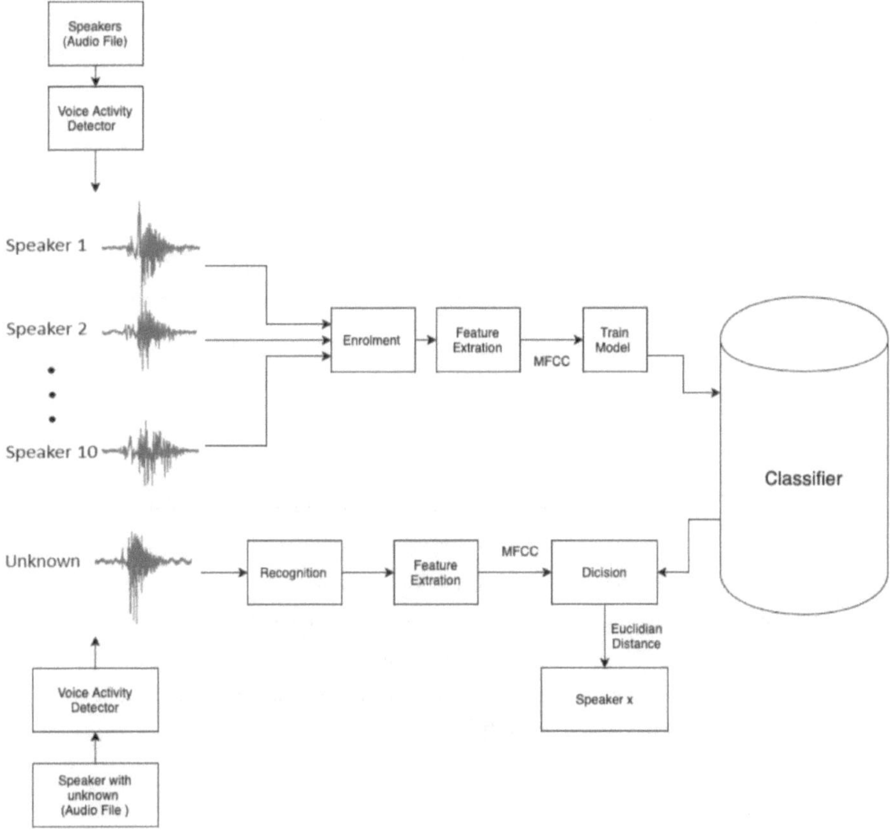

Figure 5:Speaker identification block diagram.

3.1 Vocal Activity Detection (VAD).

According to the speaker identification block diagram, since the requirement of this assignment is to identify ten speakers from through in single-track audio. The first approach is to ensure that all voice activities of the audio file which comprises ten speakers for enrollment and at least 2 for unknown are detected so that identification can take, please. In this case, the vocal activity detection (VAD) algorithm implement. Vocal activity detector work by utilizing band ratio, running and dynamic minimum, periodicity, and maximum RMS energy estimated, also include adaptive and noise resistant threshold computation, and hangover smoothing to fade in and out voice/noise boundaries and the visual representation presented through showing VAD mark sections.

```
train = 'train/trainspeaker.wav';
noise = 'noise/whitenoise_mixdown.wav';
%noise = 'noise/pink';

%With Noise

VoiceEnrolment(train, noise)
```

```
clc
train = 'speaker/test.wav';
noise = 'noise/whitenoise_mixdown.wav';
%noise = 'noise/pink';

%With Noise
VoiceRecognition(train, noise)
```

```
% Scale noise signal down
noiseSignal = 0.1 * noiseSignal;

% Mix voice and noise
inputSignal = voiceSignal + noiseSignal;
```

According to the above code, vocal activity detector perform on input audio signal and noise, though noise can be optional but it is recommended because it all easy fade in and out of each voice detected, that is when each speaker is detected and the duration that voice is silent is more visible and easy to analyze through the developed algorithm. For the input, an audio track of ten speakers are recorded and another audio track which comprises of ten speakers and at least 2 unknowns were also recorded at around 2-minute audio track and store in train/trainspeaker.wav and speaker/trainspeaker.wav respectively.

```
acfSectionalized = acf(duration+lowerLag:duration+upperLag);

% We normalizing the ACF with n - h and then energy per sample.
hVector = lowerLag:upperLag;
lagArray = duration - hVector;
acfNormalized = acfSectionalized ./ lagArray;

energyNormalized = acf(duration)/duration;

finalacf = acfNormalized / energyNormalized;

peakPeriodicity = max(finalacf);


return
```

The above code is a section of periodicity code to overcome MFCC downside noise sensitivity periodicity is introduce, this code utilizes periodicity function describe as follow.

$$P(h) = (ACF(h)/(n - h)) / (ACF(0)/n)$$

$$Tmin <= h <= Tmax$$

Where ACF is the autocorrelation function as describe in above code and the function is given by

$$ACF(h) = (1/n) \; Sum(t=1 \; to \; n - h) \; X(t+h)*X(t)$$

$$Tmin <= h <= Tmax$$

In the above code, h denotes the lag value that varies Tmin to Max, these are lower and upper limit of the pitch period respectively.

```
% Output now contains the center-clipped signal. Let's get it's ACF.
acf = xcorr(output);

% Now, we want the ACF only for lag values that fall within the pitch
% period limits. Let's take 2.5ms to 15ms, i.e. with a sampling rate of
% fs, that corresponds to:
lowerLagMS = 2.5;
upperLagMS = 15;
lowerLag = round((lowerLagMS * fs) / 1000);
upperLag = round((upperLagMS * fs) / 1000);
```

However, pitch period needed to evaluate as describe in above code, This pitch period evaluate only at the range that lie to human voice pitch (1 kHz) where the maximum of the h result to maximum p(h), therefore, this is the pitch period, in order words maximum P(h) value itself is the periodicity of the frame.

```matlab
% Noise amplification ratio compared to voice expressed as a %
noiseScalingFactor = 0.1;
```

```matlab
frameLength = 60;
frameLengthSamples = ceil((frameLength * fs)/1000);

% Segment the signal into frames
segmentedSignal = frameSplit(inputSignal, frameLengthSamples);
numFrames = size(segmentedSignal, 1);

% Initialize VAD flag arrays
iVad = zeros(numFrames, 1);
fVad = zeros(numFrames, 1);

% Initialize hang over smoothing variables
hangoverThreshold = 5;
inactiveFrameCounter = hangoverThreshold + 1;
```

The above code is where the final setting carried out, this includes noise scaling which is the amplification of noise signal compares to voice, also include the frame length at 60ms, finally the hangover threshold, this help to set a threshold for number of unvoiced frames to pad unconditionally after the end of a voice frame.

```matlab
% The buffer is full of voiced frames that need to be used
vadMarker = [vadMarker; 0.5 * ones(length(frame), 1)];
outputSignal = [outputSignal; frame'];


%Important for Speaker Identification in this project
countee = countee + 1;
if(countee > 0 ) % Lets get to voice path and save to database for Comparison
    databasefile = sprintf('traindatabase/VoiceOnlyOutput%d.wav', countee);% But File name is not used

    markvoicefile = sprintf('speakerdatabase/VoiceOnlyOutput%d.wav', countee);% But File name is not u
    audiopermark = processAudio(outputSignal);
    audiowrite(markvoicefile,audiopermark,fs);



    if ~(exist(databasefile))

        disp('Unknown Voice Detected');
```

However, video maker finally constructs to mark out all detected voice session, the above code also describes the output signal which comes from the buffer that is full of voice that needed to utilize for visual display and voice identification purpose.

3.2   Speaker Identification.

The section explains the structure of an MFCC, in this assignment, the input voice recorded at 44100Hz, the sampling frequency select to minimize the effect of aliasing in the analogue to digital conversion (Linde et al., 1980).

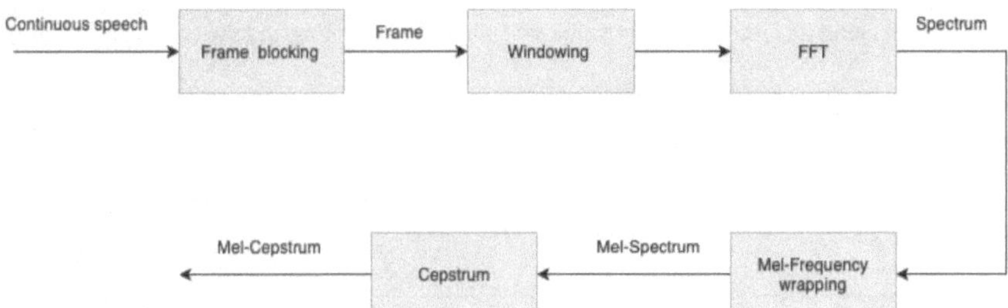

Figure 6: Implemented Block diagram of the MFC.

3.2.1   Frame Blocking:

```
function [] = test(name)


  name1 = audioread(name);

    % refer to mfcc.m file and understand the terms below and assign values as
    Tw=25;
    Ts=10;
    alpha=0.97;
    R = [300 3700];
    M = 33;
    C = 13;
    L = 22;
    hamming = @(N)(0.54-0.46*cos(2*pi*0:N-1.'/(N-1)));
    dis=zeros(1,33);
    [ tMFCCs, ~, ~ ] = mfcc( name1, 44100, Tw, Ts, alpha, hamming, R, M, C, L )
        |

    % After finding MFCC of test sample, we will find MFCC of each speaker mode
    % also Euclidian distance between the speaker model and test sample
    [speaker1,Fs]=audioread('traindatabase/VoiceOnlyOutput1.wav');
    [MFCCs1,~,~] = mfcc(speaker1, Fs, Tw, Ts, alpha, hamming, R, M, C, L );% Ca
    dis(1) = dtw(tMFCCs,MFCCs1);                                          % fi

% Framing and windowing (frames as columns)
frames = vec2frames( speech, Nw, Ns, 'cols', window, false );
```

The above code is the beginning of MFCC implementation, however, the first approach is frame blocking, in this step the continuous speech is blocked in the frame of N samples as 13 (variable C) with adjacent frames being separated by M which is 33. In this case, 5 speakers have been tested. The other 5 could be added with the same presses.

3.2.2   Widowing:

```
hamming = @(N)(0.54-0.46*cos(2*pi*[0:N-1].'/(N-1)));
```

The above code is the next step and the is window process for each individual frame to minimize the signal discontinuities at the beginning and end of each frame, the above code concept minimizes spectral distortion by using window taper at the beginning and the end of each frame.

Fast Fourier Transform (FFT):

```
nfft = 2^nextpow2( Nw );     % length of FFT analysis
K = nfft/2+1;                % length of the unique part of the FFT
```

The above code is the next step which is Fast Fourier Transform, this help to convert each frame provided by the N sample from time domain into the frequency domain. The computation time of FFT include the number of complexity reduced from N2 to (N/2)log2N as described in above code.

3.2.3   Mel-frequency Wrapping:

```
%% HANDY INLINE FUNCTION HANDLES

hz2mel = @( hz )( 1127*log(1+hz/700) );      % Hertz to mel warping function
mel2hz = @( mel )( 700*exp(mel/1127)-700 ); % mel to Hertz warping function
```

Next important approach is Mel-Frequency Wrapping; however, this follow human perception of the frequency at 1000 Hz and a subjective pitch is measure on scale called the mel as describe in above code, this mel scale is known as linear frequency below 1000 Hz and a logarithm scale space above 1000 Hz, having this result the above code formulation.

3.2.4 Cepstrum and Feature Extraction:

```matlab
%% FEATURE EXTRACTION

speech = filter( [1 -alpha], 1, speech ); % fvtool( [1 -alpha], 1 ); First order FIR Filter

% Framing and windowing (frames as columns)
frames = vec2frames( speech, Nw, Ns, 'cols', window, false );

% Magnitude spectrum computation (as column vectors)
MAG = abs( fft(frames,nfft,1) );

% Triangular filterbank with uniformly spaced filters on mel scale
H = trifbank( M, K, R, fs, hz2mel, mel2hz ); % size of H is M x K

% Filterbank application to unique part of the magnitude spectrum
FBE = H * MAG(1:K,:); % FBE( FBE<1.0 ) = 1.0; % apply mel floor

% DCT matrix computation
DCT = dctm( N, M );

% Conversion of logFBEs to cepstral coefficients through DCT
MFCC =  DCT * log( FBE );

% Cepstral lifter computation
lifter = ceplifter( N, L );
```

This is the final part of the approach MFCC as describe in above code, this include the final model of filter, windowing, magnitude of spectrum computation, triangular filter bank with uniformly spaced filter on mel space, filter bank application, the convertion of LogFBEs to cepstral and cepstral lifter computation.

3.2.5 Distance Calculation:

```matlab
dis1=sort(dis); % since we have stored all euclidian distances in an array 'dis', we will sort the array in asc
 display(dis1);
   %dis1=dis % since we have stored all euclidian distances in an array 'dis', we will sort the array in ascend
    spe={'speaker1 Detected','speaker1 Detected','speaker1 Detected','speaker2 Detected',...
        'speaker2 Detected','speaker2 Detected','speaker2 Detected','speaker2 Detected',...
        'speaker2 Detected','speaker2 Detected','speaker2 Detected','speaker2 Detected',...
        'speaker2 Detected','speaker3 Detected','speaker1 Detected','speaker3 Detected',...
        'speaker3 Detected','speaker3 Detected','speaker3 Detected','speaker3 Detected',...
        'speaker3 Detected','speaker4 Detected','speaker4 Detected','speaker4 Detected',...
        'speaker4 Detected','speaker4 Detected','speaker4 Detected','speaker4 Detected',...
        'speaker4 Detected','speaker5 Detected','speaker5 Detected','speaker5 Detected','speaker5 Detected'}; %
```

The above code is the final step for identify each speaker using Euclidian distance function and all the Euclidian distance are store in array known as 'dis', this is then sort in ascending order. Having this allow us to know the minimum Euclidian distance in order to find the speaker ID with the closest matching to the test sample.

Figure 7:Speaker identification system GUI.

The above is the GUI designed implemented concerning above algorithm and code, therefore there are two steps involved in running this GUI, this includes train of speaker, test for speaker and unknown, however, the design manual record and store 44.1kHz voice with group of people more than 7 to 12 discussing, these are two different recording audio, first audio is the once used for training, this comprises of 10 speakers, therefore for this train to carry out, use need to click on train button. Additional to that, the name and picture of the speaker would be display in the GUI axis.

The second audio is the one with an unknown voice, which also need to use as recognition to compare with the trained voice, this can be done by clicking the test button, the GUI also comprises of the edit box, this is where it will tell the user which speaker identify and when unknown voice detected.

As stated above, in order to implement such a system a several steps which were described in details in previous sections. Note that many of the above tasks are already implemented in Matlab. It comprises of noise, audio track for speakers and unknown voice, these are used to test the assignment, in this paper work each speaker extract from audio tract using two different method known as voice activities detection algorithm and final speaker identification algorithm based on MFCC feature extraction and the final identification rate can be improved by adding more vectors to the training code-word's. However, in the feature extract, Mel frequency Cepstrum coefficient was found based on human ear's critical bandwidths with frequency whereas the feature match find the vector quantized (VQ) distortion or the Euclidian distance between input utterances of an unknown speakers and this noted to store in our the database, based on the closest Euclidian distance between output of Mel frequency Cepstrum coefficient from voices signal, the final step decide which speaker identify including personal photo for the speakers also provide when unknown voice detected.

5 Simulation results.

In this section present the simulation result of speaker's identification system that was carried out using MATLAB, the result presented the graphical voice of input voice, detected voice, speakers, unknown and distance between speakers for speaker's identification purpose. In case of clarifying the above process, a four step were implement which are Train/ Enrollment Result, Recognition, GUI Result and Euclidean distance between voices as shown below.

5.1 Train/ Enrollment Result.

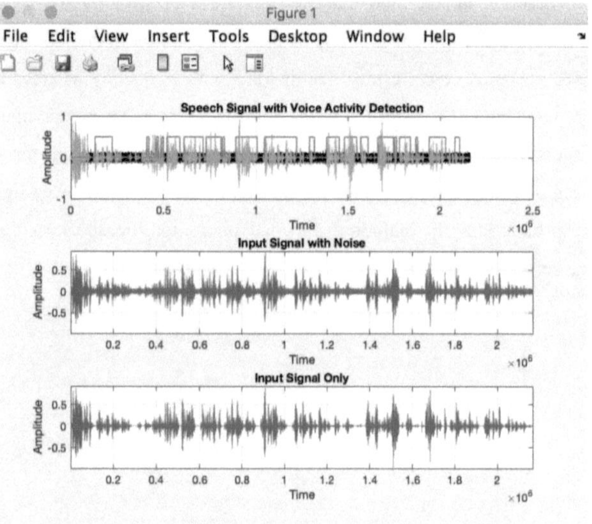

Figure 8:Trained speakers.

It can be seen in the graph above, result present the graphical view of the training session, there are three signal in this graph that utilize subplot function to plot them horizontally, these are train voice (input) which comprise of three known speakers in a single audio track as illustrated with red marker, input signal with noise, the second is the input signal mix with noise in middle, while the last below all graphs is the input signal only. According to the signal with red marker and input signal with noise, this signal is the signal that enrolled for feature extraction and speaker identification, it utilizes voice activity detection algorithm to ensure that all voice is detected and slit to train database before performing the recognition, the red marker is where all voice activity detected, the concept behind the design is that, for every voice that human speech, there is always silent, therefore the system add noise to understand the silent part and the input voice part, mark it with red color marker. Therefore, all positions where the red color marker located are the position where all voice detected and save to database for further comparison that is carried out using MFCC and distance between input voice. However, the signal with input signal with noise, this is the input signal mix with noise except that the marker is not presented in this scenario, it will notice that there is high magnitude and the silent part of the signal have heavy noise, therefore the input signal only shows the original signal voice with silent part.

5.2   Recognition.

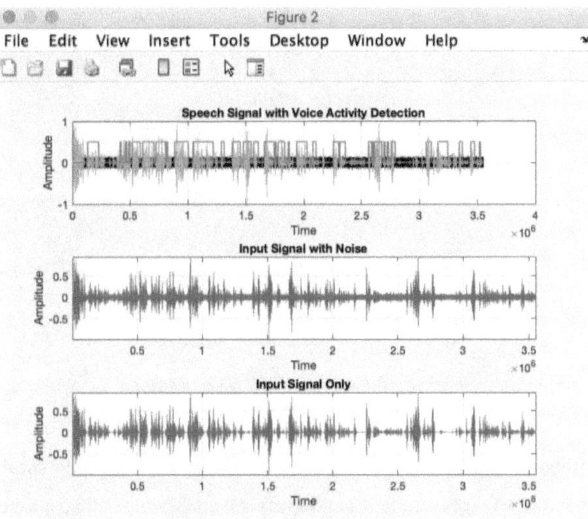

Figure 9:Identify speakers and unknown voice.

The above graph is the subplot of voice identification system, this follows similar concept of signal process as the train section explain except that, the audio track utilizes in this session have unknown speaker as it will notice that, the train audio track graph has shorter during and less marker compare to the recognition audio track presented above. According to the recognition graph above, an unknown speaker, that is the speaker that does not train starts at around 22 looking at the time and ends at approximately 35 at the time graph which makes it longer, this session is where the known speaker. Therefore, the red marker is where all the person voice involves in the audio track are split and compare with the train voice. The idea is if the voice matches the one stored in the training database, the system will display the "speaker ID" match and if there is no match, it will display "unknown speaker detected".

5.3   GUI Result.

Editor's note:
This image was removed for privacy reasons.

Figure 10:GUI speakers ID and Photo.

The GUI above present the speaker detected along with ID, three speakers trained and detected as speaker name1, speaker name2 and speaker name3 for the other speakers would follow the same procedure, therefore once user press the test button, the system would perform the simulation, this takes a few times as it processes the train audio track, compare it with recognition audio track and display the identity speaker with ID and finally display a picture for the speaker.

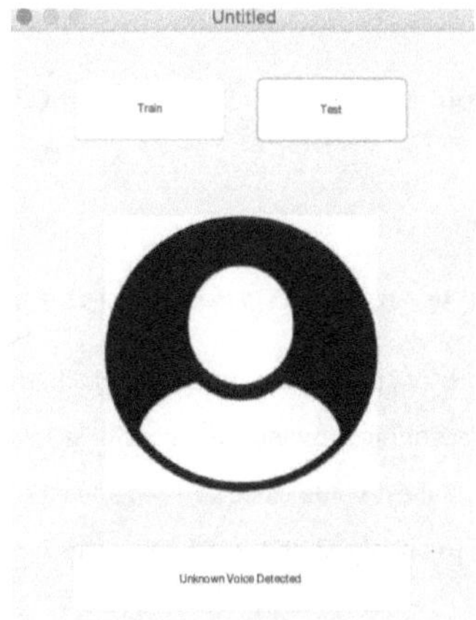

Figure 11:GUI unknown speaker.

The above GUI present the result for the un-match or unknown voice detected when comparing the train audio track with the recognition audio track, therefore as soon as unknown speaker found, the system will display known voice detected, this information the user to understand that there is unknown voice in the recognition audio track compare to the voice that has been trained and store as main speakers.

5.4   Euclidean distance between voices.

cell

'speaker2 Detected'

3.9776e+03 (Speaker 1)

1.3388e+03 (Speaker 2)

9.2690e+03   (Speaker 3)

dis1 =

1.0e+04 *

Columns 1 through 12

   0.0000   0.0730   0.0907   0.1116   0.1339   0.1754   0.3367   0.3978   0.5447   0.7276   0.7312   0.7819

Columns 13 through 20

   0.8789   0.8877   0.9152   0.9269   1.3494   1.4145   1.9486   2.0927

The above result is the sample for speaker identification which provide the closest distance to the input file, "dis1" sorted in ascending, however the result   display detected speaker2 by finding the closest distance or the smallest value close to each speaker. Therefore, the highlighted red color 1.3494 is more closer to speaker2 at 1.3388e+03. This is same method for other speaker identification.

```
Command Window
  Unknown Voice Detected
  Unknown Voice Detected
  Unknown Voice Detected
  Unknown Voice Detected
  Unknown Voice Detected
  Unknown Voice Detected
  Unknown Voice Detected
  Unknown Voice Detected
  Unknown Voice Detected
  Unknown Voice Detected
  Unknown Voice Detected
  Unknown Voice Detected
  Unknown Voice Detected
fx >>
```

Figure 12:distance between voices for unknown speakers

The result display unknown voice detected as soon as there is not closest distance to the input file, that is no voice match voice with the one store in database as enrolment or train purpose.

6   Discussion.

In this assignment, the speaker identification approach while verification does not consider, it utilizes a Mel frequency Cepstrum Coefficient (MFCC) for feature extraction in order to design the text-independent speaker identification system. The extracted speech feature of the speaker was quantized to the number of centroids with the help of the vector quantization algorithm concept, this method comprises of a code block of that speaker, therefore MFCC is calculated during the training phase and tested in the testing phase. Voice activity detected the people in training sessions and tested later in the testing session. However, the Euclidean distance between the MFCC's of each speaker during the training phase to the centroids of the individual speaker at the testing phase was all measured and the speaker is identified according to the minimum Euclidean distance. Feature extraction involves finding MFCCs of all input voice or speech signal, therefore the MATLAB programming code consider windowing, Mel Spaced Filter Banks and convert the speech signal to a parametric which is also can call Mel Frequency Cepstrum Coefficients, in this code, the algorithm MFCC are based on known variation of human ear's frequency at 1 kHz. During the feature matching, the input utterance of the signal that has an unknown speaker is converted to MFCCs while total distortion between these MFCCs and the codebooks or train stored in the database were measured. Based on vector quantization distortion, the system able to decide the speakers from the unknown speaker.

The reason of selecting the MFCC that it is simple to follow and design and use physical or practical scenario such as frequency for features extraction, however, it also provide a low correlation between coefficients, compares to LPC which is highly correlated. MFCC algorithm similar to the human auditory perception system and important phonetics characteristics can easily caption while LPC can be complicated to capture.

7 Conclusion.

In conclusion, the code developed MATLAB environment and performs for speaker identification system using MFCC feature extraction and voice activities detection algorithm is satisfactory. The system utilizes different recorded signals for speech train and recognition to assure optimum performance of the identification system and the ID of all speaker identity are presented. The work is applied in many areas which are Access Control Systems, Time and Attendance Systems, Voice command and control, Telephone-Banking/Booking, Biometric System and Security Control System for confidential information such as for Forensic purposes.

8 References.

Deller J.R., Hansen J.H.L. & Proakis J.G. (2000). Discrete-Time Processing of Speech Signals. *IEEE Press*.

Fu Zhonghua and Zhao Rongchun. (2003). An overview of modeling technology of speaker recognition. *IEEE Proceedings of the International Conference on Neural Networks and Signal Processing*.

Gaurav Giroti, Tejas Nakhate, Mahesh Laddha, Prof. Manoj Sarve. (2018, May). Person Identification through Voice using MFCC and Multi-class SVM. *International Research Journal of Engineering and Technology (IRJET), 5*(5), 690-693.

Hagen A., Connors D.A. & Pellm B.L. (2015). The Analysis and Design of Architecture Systems for Speech Recognition on Modern Handheld-Computing Devices.

Ishizuka K. and Nakatani T. (2016). A feature extraction method using subband based periodicity and aperiodicity decomposition with noise robust frontend processing for automatic speech recognition. *Speech Communication*.

Motlíček P. (2002). Feature Extraction in Speech Coding and Recognition. *Oregon Graduate Institute of Science and Technology*, 1-50.

Pialy Barua, Ainul Anam Shahjamal Khan, Muhammad Sanaullah. (2014). Neural Network Based Recognition of Speech Using.

Sanjay A. Valaki, Harikrishna B. Jethva. (2016). A Survey on Feature Extraction and Classification Techniques for Speech Recognition. *IJARIIE, 2*(6), 830-837.

Urmila Shrawankar. (2013). Techniques for Feature Extraction in Speech Recognition System: A Comparative Study. *ResearchGate*.

Vibha Tiwari. (2010). MFCC and its applications in speaker recognition. *International Journal on Emerging Technologies*, 19-22.

Y. Linde, A. Buzo & R. Gray. (1980). An algorithm for vector quantizer design. *IEEE Transactions on Communications*, 84-95.

Zilovic, M.S.; Ramachandran, R.P.; Mammone, R.J. (2014). Speaker identification based on the use of robust cepstral features obtained from pole-zero transfer functions. *IEEE Transactions on Speech and Audio Processing*.